EXPERIMENTS
IN STRENGTH OF MATERIALS

EXPERIMENTS
IN STRENGTH OF MATERIALS

BELA I. SANDOR
University of Wisconsin

PRENTICE-HALL, INC., *Englewood Cliffs, New Jersey 07632*

Library of Congress Cataloging in Publication Data

SANDOR, BELA IMRE.
    Experiments in strength of materials.

    1. Strength of materials—Laboratory manuals.
    I. Title.
TZ412.S26   620.1'12'028   79-22647
ISBN 0-13-295329-3

Editorial/production supervision and interior design: Nancy Moskowitz
Cover design: A Good Thing, Inc.
Manufacturing buyer: Gordon Osbourne

Printed in the United States of America

10   9   8   7   6   5   4   3   2   1

PRENTICE-HALL INTERNATIONAL, INC., *London*
PRENTICE-HALL OF AUSTRALIA PTY. LIMITED, *Sydney*
PRENTICE-HALL OF CANADA, LTD., *Toronto*
PRENTICE-HALL OF INDIA PRIVATE LIMITED, *New Delhi*
PRENTICE-HALL OF JAPAN, INC., *Tokyo*
PRENTICE-HALL OF SOUTHEAST ASIA PTE. LTD., *Singapore*
WHITEHALL BOOKS LIMITED, *Wellington, New Zealand*

CONTENTS

PREFACE

Experimental work in the area of strength of materials can be very useful to students for several reasons. First of all, it enhances their knowledge of the fundamental concepts involved. This includes realistic assumptions and approximations in theoretical work, the nature and ranges of common errors, and the limitations of theories and basic formulas. The development of interest through personal involvement is helpful in the learning process.

Experiments also show that even simple projects may present significant challenges and difficulties. The appreciation of this is important for all engineers and scientists whose ideas are normally put into practice by others. Laboratory work often encourages some cooperation with other people, another feature of professional work after graduation.

The projects presented here are in two categories: independent projects that require very little equipment, and experiments that are most efficient when performed by a technician or instructor for a whole class of students. The latter are generally those that require complex equipment or specimens in limited supply. The primary use of each project is shown.

The order of performing experiments is difficult to decide when the students have not yet had a full course of lectures on the subject. In this case, some experiments may have to be done before all of their concepts are presented and discussed in lectures. Yet, the students may still benefit from the experiments.

The projects may be handled in a flexible way, doing more or less what is described, or making desirable alterations to suit local conditions. Basic experiments for whole classes can be repeated semester after semester, while simple independent projects may be rotated freely.

Finally, some valuable advice to ambitious and hard working students in seeking employment. Employers generally prefer to hire those technical people who have sound knowledge of the fundamentals and also have experience with the practicalities of their professions. Furthermore, managers want engineers who can work with a minimum of supervision and can communicate the results efficiently and accurately (obviously, the best of these can also eventually become managers). My advice, on the basis of numerous success stories of this kind, is that the student gradually build a portfolio of reports on independent laboratory projects, and proudly present it during interviews. Perfection is not important, the existence of such a portfolio is. This set of projects offers numerous opportunities for independent work to satisfy many different interests. Good luck!

Bela I. Sandor

Madison, Wisconsin

EXPERIMENTS
IN STRENGTH OF MATERIALS

1

TENSION TEST

BACKGROUND

All solid materials deform when forces are applied to them. The deformation depends on the magnitude of the force, the chemistry and condition of the material, the specimen geometry, temperature, loading rate, etc. Force versus deformation curves can provide useful information about the mechanical behavior of a given specimen. However, the information is not readily applicable to specimens with different geometries, even if the material is the same. The most useful information is from stress versus strain plots which are independent of specimen geometry (unnotched specimens).

OBJECT

Perform an axial tension test on an unnotched steel or aluminum specimen (ASTM standard or similar size). Collect stress and strain data throughout the test, all the way to fracture and analyze it as much as possible. Later, return to analyzing the data when new knowledge allows you to obtain further information.

EQUIPMENT

1. Mechanical testing machine.

2. Load and deformation measuring devices.

3. Micrometer.

EXPERIMENT

1. Perform the tension test.
2. Record load vs. deformation throughout the test.
3. Record load vs. minimum cross-sectional area throughout the test (optional).

REPORT

1. Describe the equipment and test procedures.
2. Plot the engineering and true stress-strain curves on the same diagram (use arithmetic coordinates).
3. Determine (as soon as you understand the concepts):
   (a) Elastic modulus, E
   (b) Proportional limit
   (c) Yield strength at 0.2% offset, σ_y
   (d) Ultimate strength, σ_u
   (e) Modulus of resilience
   (f) Percent reduction of area
   (g) Percent elongation over the given gage length
   (h) True toughness
   (i) True fracture strength, σ_f
   (j) Fracture ductility, ϵ_f
   (k) Strain hardening exponent, n (from $\sigma = K\epsilon_p^n$)
   (l) Strength coefficient, K
4. Describe your observations of changes in the specimen during the test.

TENSILE TEST OF _____

General Data: Diameter of specimen _____ in.
Area _____ in.²
Type of Extensometer _____
Gage Length of Extensometer_____ in.

Load lb	Extensometer reading in.	Increment in reading in.	Strain in./in.	Stress psi	Remarks
O		O	O	O	

Data taken after failure:

Length over gage pts. ___ in.   Elongation _____%
Neck diameter   ___ in.   Reduction in area_____%

Character of fracture _____

3

2

TENSION TEST OF PLEXIGLAS

BACKGROUND

Tension tests of brittle materials require special consideration because such materials are sensitive to stress concentrations. Any discontinuities in a member, even small flaws and invisible scratches, may cause failure at relatively low load.

OBJECT

Perform tension tests of smooth and scratched flat bars of plexiglas.

EQUIPMENT

  1. Mechanical testing machine.

  2. Micrometer.

  3. Load measuring device.

  4. Extensometer (optional).

EXPERIMENT

  1. Cut smooth specimens from flat plate with 2 in. (5 cm) gage length. Shape the specimen ends as shown in Figure 1(a). Polish the curved and straight test section of each specimen to eliminate scratches.

2. Make steel grips as shown in Figure 1(b). Figure 1(a) must easily fit in Figure 1(b). The curved parts must be matched to minimize the contact stresses during loading.

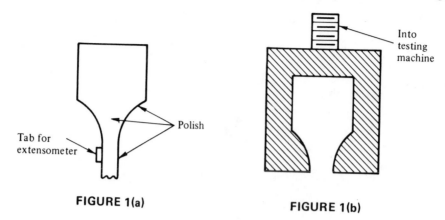

FIGURE 1(a) FIGURE 1(b)

3. If an extensometer is used, glue tabs to the specimen so that the extensometer may be attached to the specimen without scratching the latter.

4. Perform the tension test and record stress and strain data.

5. Observe the optical changes in Figure 1(a) where it contacts the grips.

6. Make a minor circumferential scratch at the middle of a new specimen. The change in cross-sectional area should be small and known to within a small percentage of the original area.

7. Perform the tension test as before.

REPORT

1. Describe the specimen, grips, other equipment, and test procedures.

2. Plot stress-strain curves and determine as many properties as possible according to the preceding "Tension Test" requirements.

3. Describe the optical changes in the specimens during the tests.

4. Compare the fractures to those of ductile metals.

5. Something is seriously wrong with the stress-strain plot for the scratched specimen. What is it?

3

TENSION TEST OF WOOD

BACKGROUND

The fibrous composition of wood makes its tensile and shear strengths highly dependent on loading orientation with respect to the fibers. For example, the gripping arrangement described in "Tension Test of Plexiglas" is not perfect for wood with longitudinal fibers because of possible splitting along those fibers.

OBJECT

Perform tension tests of wood loaded parallel to the fibers. In separate tests, use loading perpendicular to the fibers.

EQUIPMENT

Same as for plexiglas (Exp. 2).

EXPERIMENT

1. Develop modifications of the gripping used for plexiglas, or adopt a new technique, for gripping wood.

2. Perform tension tests on smooth specimens for both fiber orientations. Record stress and strain data. Tabs are not necessary for the extensometer.

REPORT

1. Describe the wood, gripping, and test procedures.

2. Plot stress-strain curves and determine as many properties as possible according to the basic "Tension Test" requirements.

4

THIN FIBERS IN TENSION

BACKGROUND

Very thin fibers of metals, glass, polymers, etc. are used as major load-carrying components in modern composite materials. Tensile tests of such fibers require special considerations and can be quite challenging compared to tests of common engineering materials.

OBJECT

Test at least one kind of fiber, as thin as possible. For a partial list of possibilities, consider metals, glass, carbon, kevlar, nylon, dacron, rayon, hair, silk, boron, wool. Determine the complete stress-strain curve and investigate the mode of fracture.

EQUIPMENT

1. Dead weights or a mechanical testing machine with a sensitive load cell.

2. Deformation measuring device that applies little or no force (optical) to the fiber specimens.

3. Optical microscope (also scanning electron microscope if available).

4. Micrometer of high enough resolution to measure the diameter of the fibers. This may not be possible for very thin fibers; use a microscope in that case.

EXPERIMENT

1. Develop a method of gripping the ends of the test pieces. Wrapping the fibers on mandrels or gluing tabs on them may be helpful.

2. Check the fibers for microscopic flaws.

3. Determine the diameters of the test pieces and the variation of these along the test sections.

4. Test the fibers and plot the stress-strain curves.

5. Measure the diameter of the fracture area.

6. Analyze the fracture surfaces with a microscope.

REPORT

1. Describe the details of the experimental setup and procedures.

2. State the results and the range of experimental error. Include the percent reduction of area.

3. Describe the features of the fracture surfaces and theorize about the mode of fracture (ductile or brittle).

5

TRUE STRESS AND TRUE STRAIN

BACKGROUND

In ductile metals large deformations occur when loading causes gross yielding. These deformations must be considered in determining the true stresses and strains in some members. In addition, they affect some important material properties, namely the true fracture strength, fracture ductility, and strain hardening exponent.

OBJECT

Test a ductile metal in tension and compression and obtain true stress and strain data. From a plot of these, determine the material properties mentioned above and the power function that describes the stress-strain behavior in the plastic range of deformations.

EQUIPMENT

1. Mechanical testing machine with load measuring device.
2. Micrometers: flat-tipped, and point- or round-tipped.

EXPERIMENT

1. Use round bars of mild steel or aluminum with a diameter of about 1 to 1.5 cm. The length of the tension specimen should allow a free length of 15 to 20 cm between

the grips. Make a gage mark every 2 cm along the free length (do not notch the specimen). The compression specimen should be twice as long as its width.

2. Perform the tests. Make sure that the diameters recorded are always the minimum values in the tension test (use a round-tipped micrometer), and the maximum values in the compression test (flat-tipped micrometer).

3. Immediately after fracture in tension, run your finger along the length of the specimen to feel the differences in temperature that resulted from the plastic deformation.

REPORT

1. Describe the equipment and test procedures.

2. Calculate and plot true stress versus true strain for tension and compression on log paper.

3. Determine the true fracture strength, fracture ductility, strength coefficient, and strain hardening exponent from the plot of the tension data.

4. Give the stress-strain equation for the material tested. Consider the plastic range only.

5. Check the data to see if the maximum load in tension is reached when strain equals strain hardening exponent.

6. Calculate the percent elongation using as many different gage lengths as possible. Compare the results and decide whether percent elongation is a good material property. Discuss your findings and reasoning.

DATA FOR TRUE STRESS – TRUE STRAIN PLOT

TENSION

Dia. in.	Area in^2	A_O/A	$\ln \dfrac{A_O}{A} = \in$	$\dfrac{P}{A} = \sigma$, psi	Load lb
≈.500	≈.196			O	O
.495	.192				
.490	.189				
.485	.185				
.480	.181				
.475	.177				
.470	.173				
.460	.166				
.450	.159				
.440	.152				
.430	.145				
.420	.139				
.410	.132				
.400	.126				
.390	.119				
.380	.113				
.370	.107				
.360	.102				
.350	.096				
.340	.091				
.330	.086				
.320	.080				
.310	.075				
.300	.071				

COMPRESSION

Dia. in.	Area in^2	A_O/A	$\ln \dfrac{A_O}{A} = \in$	$\dfrac{P}{A} = \sigma$, psi	Load lb
				O	0
					10000
					11000
					12000
					14000
					16000
					18000
					20000
					22000
					25000
					28000
					32000
					36000

6

BIAXIAL TENSION AND COMPRESSION

BACKGROUND

The ability of a metal to deform plastically depends on its intrinsic ductility and the state of stress inside it. The magnitudes and signs of the principal normal stresses are important because the absolute maximum shear stress (the stress causing plastic flow) depends on them. Biaxial loading may increase a metal's ability to yield, in comparison with uniaxial loading.

OBJECT

Compare the behaviors of mild steel tested in uniaxial tension and in biaxial tension and compression.

EQUIPMENT

1. Mechanical testing machine with load measuring device. Deformation measuring device is helpful but optional.

2. Micrometer or vernier calipers.

EXPERIMENT

1. Pull in uniaxial tension a bar of hot-rolled mild steel. The bar should have a rectangular cross section, not more than 1 cm on either side. Rub the test section of the bar with chalk dust. Record the yield, ultimate, and fracture loads.

2. Pull an identical steel bar which has a clamp compressing it in the middle. The contacts of the clamp should not be too sharp or too wide; about a 0.5 cm^2 contact area is reasonable. Rub the test section of the bar with chalk dust. Record the yield, ultimate, and fracture strengths.

REPORT

1. Describe the specimens and the test setup.

2. Calculate the yield, ultimate, and fracture stresses and the percent reductions of areas.

3. Discuss your observations of the progressive yielding in each test.

4. Make an attempt to calculate the compressive stress caused by the clamp.

7

COMPRESSION TESTS

BACKGROUND

Compression tests are necessary as well as tension tests to characterize completely the behavior of a material under static axial loads. Generally, they are simpler than tension tests, but safety precautions must be taken when the material is brittle (shattering).

OBJECT

Perform compression tests on smooth cylindrical or prismatic specimens of several materials.

EQUIPMENT

1. Mechanical testing machine with compression platens or grips.

2. Vernier calipers.

3. Load measuring device.

4. Short steel pipe to surround the specimens during test.

EXPERIMENT

1. Make short, smooth specimens with length approximately twice the diameter. Suggested materials are mild steel, aluminum, cast iron, plexiglas, epoxy, plaster, wood (various grain orientations), and fiberglass.

2. Do not use extensometers since they are easily smashed.

3. Perform the compression test and record the yield strength (if observable) and the ultimate strength.

REPORT

1. Describe the specimen, test equipment, and procedures.

2. Give the initial and final dimensions of the specimen.

3. State the engineering and true strength values.

4. Sketch the specimen after failure.

8

COMPRESSION TEST OF CONCRETE

BACKGROUND

The properties of concrete depend on the proportions of fine and coarse aggregate, water, and cement used in making it. The workability of a mix strongly depends on the amount of water used. It is measured by the so-called *slump test* (the slump of a truncated cone of concrete mix after its metal form is lifted up). The mixture hardens by a chemical reaction called *hydration*. This occurs rapidly at first, but continues at a decreasing rate for an indefinite period.

OBJECT

Make concrete cylinders of identical dimensions but with different proportions of constituents. Test the cylinders in compression.

EQUIPMENT

1. Molds and tools for making concrete specimens.

2. Mechanical testing machine with load cell and compression platens.

EXPERIMENT

1. Make the concrete cylinders under expert guidance. Three different mixes of concrete are generally sufficient. It would be interesting to have at least one specimen with a helical reinforcing wire in it. Record data as soon as possible on the data sheet.

2. Test the cylinders in compression after about four weeks of curing.

REPORT

1. Evaluate your specimens with respect to the table which shows practical uses for various mixes of concrete.

2. Estimate the reasons for the differences in strength.

USES FOR VARIOUS MIXES OF CONCRETE

General properties of concrete and its main uses	Ultimate strength σ_u at 28 days	Weight basis W''/C ratio
High early strength; pavements, columns, beams, structures severely exposed	5500 psi	0.40
High strength, watertight; dams, piers, walls, pavements; structures severely exposed	4500 psi	0.50
Medium strength, watertight, frost resistant; dams, piers, walls, pipes, pavements	3800 psi	0.55
Medium strength, moderate exposure (southern climate); dams, piers, walls	3000 psi	0.60
Low strength, protected environment (not for severe exposure or wetting and drying cycles); foundations, enclosed walls	2200 psi	0.65
Low strength, protected environment; used in bulk	1800 psi	0.75

SAMPLE DATA SHEET

Weight Per Cubic Foot Test Data

Material	Volume in ft^3	Material & Container	Container	Per ft^3
		Weight in Pounds		
Sand	0.1	16.4	5.2	112
Gravel	0.5	74.4	19.4	110
Concrete				
Group 1				
Group 2	0.5	94.75	19.4	150.7
Group 3				

Moisture Absorption Data

Madison Sd. Absorbs _0.4_ % Water/lb

Madison Gl. Absorbs _1.5_ % Water/lb

Starting Batches

Group 1 Pounds		Group 2 Pounds		Group 3 Pounds	
Sand	26	Sand	37+5	Sand	42
Cement	28	Cement	14	Cement	9
Gravel	46	Gravel	52+3	Gravel	47
Water	13	Water	9.5	Water	9.5

Data on Batches

	Group 1	Group 2	Group 3
Weight of cement (C), lb		14	
Total weight of fine aggregate (sand), lb		42	
Total weight of coarse aggregate (gravel), lb		55	
Total weight of aggregate, lb		97	
Total weight of water (W), lb		9.5	
Water absorbed by aggregate (W′), lb*		≈1	
Weight of free water (W″=W−W′), lb**		8.5	
Total weight of all components, lb		120.5	
Density of wet mix, lb/ft^3		150.7	
W″/C by weight		0.608	
Proportions by weight (cement/ sand/ gravel)		0.12/0.35/0.46	
Observed slump, in.		1.25	
Average ultimate strength, psi			

* W′ = (42 lb sand) (0.004) + (55 lb gravel) (0.015) ≈ 1
** W″ is available for reaction with cement

Groups 1 and 3 makes 2 cylinders and 1 beam;
Group 2 make 3 cylinders (one with spiral wire)

Cylinders: 6 in. dia. × 12 in. long
Beams: 2 in. × 4 in. × 36 in. Test over 34 in. span
One 3/8 in. reinforcing steel bar, 1 3/16 in. from
bottom of beam; σ_y = 50,000 psi

Date Made _____ Date Tested _____

9

POISSON RATIO OF RUBBER

BACKGROUND

The mechanical behavior of polymers is distinctly different from that of metals. Although the explanation of some phenomena in polymers is beyond the scope of elementary mechanics, the basic experimental techniques are still valid for different classes of solids. In some cases, such as trying to observe the Poisson effect, the experiments with polymers are simpler than with metals.

OBJECT

Measure the Poisson ratio or rubber in a simple experiment. Compare the result with known values for several metals.

EQUIPMENT

1. Small weights.

2. Flat bar for loading arm.

3. Ruler, micrometer, or vernier calipers.

EXPERIMENT

1. Take a rectangular eraser or a similar block of soft rubber. Mark four points in a square pattern on the block as shown in Figure 2 and measure the distances AB and CD when the block is not loaded.

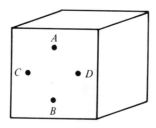

FIGURE 2

2. Apply a small load with the aid of the loading arm (this is for stability) as shown in Figure 3. Determine the strains in AB and CD.

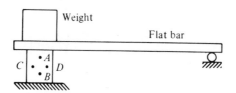

FIGURE 3

3. Determine the strains in AB and CD using at least one other weight.

REPORT

1. Describe the specimen and the test procedure.

2. Calculate the Poisson ratio.

3. Compare the result with known values for a wide range of metals.

4. *Optional:* explain the behavior of rubber on the basis of your independent reading or discussions with experts.

10

BEAM STIFFNESS-TO-WEIGHT RATIO

BACKGROUND

Often, structural members must be stiff to limit their elastic deformations, even when the danger of fracture or permanent (plastic) deformation is remote. For example, airplane wings must not touch the ground during landing even if their deformations are entirely elastic.

OBJECT

Determine the stiffness-to-weight ratio of one or more simply supported homogeneous beams loaded with a concentrated center load.

EQUIPMENT

1. Wood, aluminum, or steel bars. Cross section 1 in. X 1 in. (2.5 cm X 2.5 cm).

2. Weights or mechanical testing machine.

3. Dial gage or other sensitive deflection measuring device.

4. Weight scale.

EXPERIMENT

1. Set the supports for a span of $L = 20$ in. (50 cm).

2. Measure the deflection where the load is applied as shown in Figure 4. Make sure that the supports or the dial gage probe do not dig into the beam.

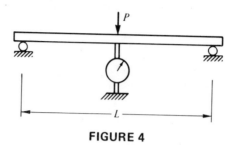

FIGURE 4

3. Measure the deflections for 5 to 10 small loads (avoid inelastic deformations).

4. Determine the weight W of the beam in the length L (the length of overhang is not important).

REPORT

1. Describe the experimental setup and procedure.

2. Plot the data as shown in Figure 5 and determine the slope k (spring constant) for the data. If there is a curve, draw a tangent to get the initial slope.

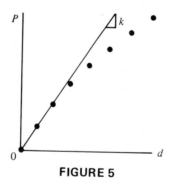

FIGURE 5

3. Calculate the stiffness-to-weight ratio, $\frac{k}{W}$.

4. If different materials were used, compare the results and discuss their practical implications.

26

11

SKIS IN BENDING

BACKGROUND

The stiffness-to-weight ratios of skis should be of interest to all skiers who have knowledge of engineering mechanics. Such information is easily obtained experimentally, so the challenge is in the interpretation of the results and in their practical use.

OBJECT

Perform three-point bending tests on several different skis that are similar in length. Evaluate the results with respect to their practical significance.

EQUIPMENT

1. Mechanical testing machine or weights.

2. Dial gage or ruler.

3. Scale for weighing.

EXPERIMENT

1. Weigh each ski available for testing.

2. Measure load versus deflection (a few cm maximum) for each ski, using the same span between supports.

REPORT

1. Describe the skis tested and the test setup.
2. Plot load versus deflection as in the "Beam Stiffness-to-Weight Ratio" experiment.
3. Calculate the stiffness-to-weight ratio, k/W, from the plot of P versus d.
4. Compare the results and discuss their practical implications, if any.

12

BEAM LOAD CELL, OPTICAL

BACKGROUND

Optical means of indicating deformations offer unique opportunities for obtaining high resolution (sensitivity). Another advantage, in some cases, is that the measurements can be made with a minimum of physical interference with the member.

OBJECT

Construct a cantilever beam load cell with an optical indicating system. Increase the sensitivity of the device as much as possible while maintaining its reliability for repeated use.

EQUIPMENT

1. Beam with small mirror.

2. Light source.

3. Weights.

EXPERIMENT

1. Set up the equipment as shown in Figure 6, to reflect the light on the wall.

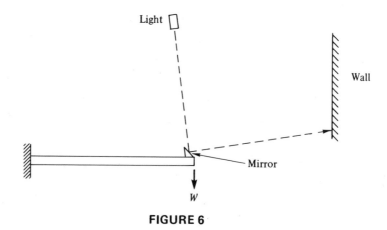

FIGURE 6

2. Make refinements to increase the sensitivity for distinguishing very small changes in the load.

3. Calibrate the load cell by plotting W versus displacement of the light spot on the wall.

REPORT

1. Describe the details of the final device and its arrangement.

2. Discuss the linearity, sensitivity (resolution), and accuracy (repeatability) of the load cell.

3. Discuss the prospects for further improving the linearity, sensitivity, and accuracy of your device.

13

STRENGTH OF GLASS RODS

BACKGROUND

Even small notches, scratches, and other defects can be detrimental to the strength of brittle members. Size and environmental conditions may also affect the strength of the member, although not necessarily the true fracture strength of the material (which is a material property and does not depend on geometry).

OBJECT

Perform bending tests on dry and wet glass rods of various sizes and surface conditions to determine the causes of high and low strength.

EQUIPMENT

1. Loading fixture for four-point bending.
2. Dead weights or sensitive testing machine to apply small loads in a controlled fashion.
3. Safety glasses.
4. Microscope of low magnification (optional).

31

EXPERIMENT

1. Carefully observe the surfaces of all glass rods to be tested.

2. Fracture several rods of different diameters in four-point bending. Use dry and wet rods in each size. Determine the apparent fracture stress for each specimen.

3. Etch several specimens with hydrofluoric acid, rinse, and test them in the dry or wet condition.

REPORT

1. Describe the test procedures, specimen sizes, and the initial surface conditions of the specimens.

2. Compare the test results and explain the observed behavior.

3. Make predictions concerning the strength of glass rods identical in material but smaller or larger in diameter compared to those used in the tests.

14

ICE BEAMS

BACKGROUND

The bending strength of ice is of great practical significance, mainly with respect to ice-breaking techniques used in harbors and water ways. Interestingly, ice is not a homogeneous material, and its strength depends on many factors such as age, temperature, and salt content.

OBJECT

Perform bending tests of simply supported ice beams under various conditions.

EQUIPMENT

1. Freezer.

2. Dead weight loading facility or mechanical testing machine.

3. Dial gage or other extensometer (optional).

EXPERIMENT

1. Make identical ice beams from fresh water and salt water.

2. Test the beams in three- or four-point bending. If possible, use beams of widely different ages and temperatures.

3. Record the load and deflection at fracture. Be careful not to smash the extensometer.

REPORT

1. Describe the specimens and the test setup.

2. Plot the fracture load versus the individual parameters selected on separate diagrams.

3. Make a master diagram of load versus deflection at fracture including all of the data. Compare the results for various beams and draw your conclusions on the basis of potential practical significance.

15

BENDING TEST OF CONCRETE

BACKGROUND

Concrete has practically no strength to resist tension. For this reason, concrete beams to be used in flexure must have steel reinforcing rods on the side in tension. Otherwise, the basic concepts of proportioning the concrete mixes are applicable.

OBJECT

Make reinforced concrete beams with rectangular cross sections. Test these in three-point bending.

EQUIPMENT

1. Molds and tools for making the beams.
2. Mechanical testing machine with load cell.

EXPERIMENT

1. Make the beams under expert guidance. It is best to use the same concrete mixes as used for the compression specimens (Exp. 8).

2. Test the beams in three-point bending after about four weeks of curing. Be careful not to apply highly concentrated loads anywhere on the beams. For example, use flat steel plates at the supports to distribute the local loads.

REPORT

1. For each beam, state the load at which the first cracks appeared.
2. Calculate the maximum compressive and tensile stresses in the concrete at the onset of cracking.

16

FLEXURE OF I BEAM

BACKGROUND

It is easy to demonstrate experimentally, with a number of strain gages mounted on a beam, that the assumptions made in developing the flexure formula are correct. The same experiment shows that the theoretical and experimental stress analyses at various points in the beam provide essentially identical results.

OBJECT

Determine the stresses theoretically and experimentally at numerous locations in an I beam that is deformed elastically only.

EQUIPMENT

1. Mechanical testing machine.

2. Load and deflection measuring devices.

3. I beam with strain gages (see Figure 7 for recommended gage locations).

4. Strain readout or recording device.

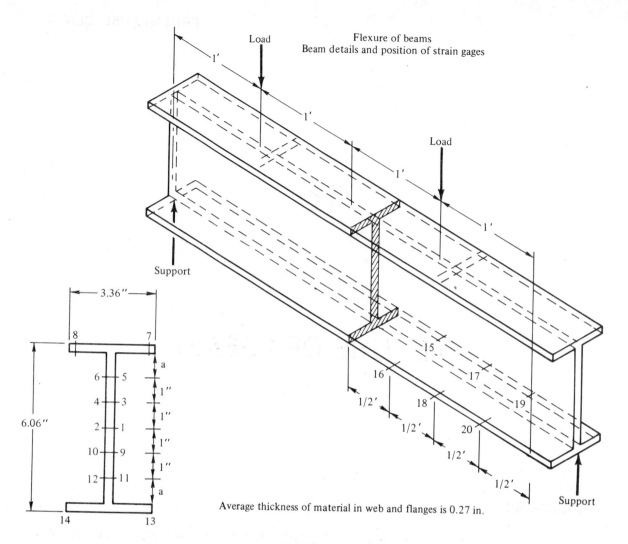

Flexure of beams
Beam details and position of strain gages

Average thickness of material in web and flanges is 0.27 in.

FIGURE 7

EXPERIMENT

1. Load the simply supported beam in four-point bending.
2. Record the strains at all gage locations and the center deflection.

REPORT

1. Compute the stress at each gage location from the measured strain and the elastic modulus of the beam.

2. Compare the experimentally determined stresses with the ones obtained using the flexure formula.

3. Plot the measured stress distribution for the center cross section. Locate the neutral axis.

4. Calculate the maximum deflection and compare it with the measured deflection.

FLEXURE DATA

Strain indicator equipment _____

Initial strain reading _____

Load on specimen _____

Gage No.	Strain reading	Strain increment	Actual strain	Stress from strain	Stress from Mc/I
1					
2					
3					
4					
5					
6					
7					
8					
9					
10					
11					
12					
13					
14					
15					
16					
17					
18					
19					
20					

17

MAXIMUM DEFLECTION IN A BEAM

BACKGROUND

The location of maximum deflection in some beams is obvious (for example, symmetrically loaded beams). If it is not obvious, it can be found fairly accurately simply by guessing However, the magnitude of the deflection is more difficult to guess. A reasonable sketch of the expectable deflection curve (the shape of the beam under load) shows the region of largest deflection. It is interesting to compare the results from guessing, experiments, and calculation.

OBJECT

Guess the location and magnitude of the maximum deflection of a simply supported, eccentrically loaded beam before performing any tests. Then check the accuracy of the guesses by experiments.

EQUIPMENT

1. Any beam of constant cross section and material that can be conveniently loaded to have measureable deflections.

2. Weights or mechanical testing machine.

3. Dial gage or other deflection measuring device.

ESTIMATE OF DEFLECTION

Estimate the location and magnitude of maximum deflection for the given beam and loading that will be used later in the experiment. Assume the load is at $0.2L$ from either support (L = distance between supports).

EXPERIMENT

Determine the location and magnitude of the maximum deflection using the above assumed load.

CALCULATION

Calculate the location and magnitude of the maximum deflection on the basis that the slope of the beam is zero where the deflection is maximum.

REPORT

1. State the results.
2. Determine the magnitudes of the differences found using the three methods.
3. Discuss the general accuracy expected when making guesses of beam deflections.

18

BEAM WITH NAILS

BACKGROUND

Nails are commonly driven into wood without any consideration for possible changes in the strength of the wood. Potentially, nails can alter the effective stiffness and the fracture strength of wood.

OBJECT

Determine the effect of nails on the stiffness and fracture strength of small wood beams. As a minor objective, also determine the behaviors of beams of different fiber orientations.

EQUIPMENT

1. Dead weights or a mechanical testing machine to apply three-point bending.

2. Ruler, dial gage, or extensometer to measure deflections.

EXPERIMENT

1. Design and cut small rectangular beams that you will be able to break with the available equipment.

2. Test the control samples without nails as in Figure 8. Plot the load P versus the maximum deflection d, all the way to fracture. Be careful not to smash the deflection measuring device.

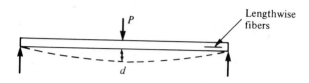

FIGURE 8

3. Test beams with a single row of identical nails on the tension side of each beam, under the load P as in Figure 9. Again plot P versus d.

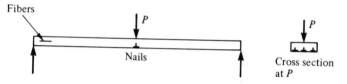

FIGURE 9

4. If you test for fiber orientation, repeat steps 2 and 3 for the beam shown in Figure 10 (overall geometry is always the same).

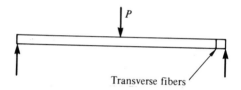

FIGURE 10

REPORT

1. Describe the details of the experimental setup and procedures.

2. State the results and the range of experimental error.

3. Theorize about the results to explain the observations.

4. Discuss how the results might change if the nails were sharper or blunter than the ones used.

19

SUPER BEAMS

BACKGROUND

Sometimes it is necessary to make beams with very high strength- or stiffness-to-weight ratios. In some cases, the cost of construction is relatively unimportant if the beam is superior to all others in its class.

OBJECT

Design and construct a beam to maximize its stiffness-to-weight ratio. The experiment is similar to the basic "Beam Stiffness-to-Weight Ratio" experiment for homogeneous beams (Exp. 10).

RULES

1. You may use any material or construction technique available.

2. Any geometry of the beam is acceptable as long as the beam (without load or supports) could be put inside a long hole with a square cross section, 1 in. X 1 in. (2.5 cm X 2.5 cm).

3. The deflection must be measured at the point where the load is applied, or be identical to that deflection.

4. The beam must carry the applied load by itself.

REPORT

1. Describe the principles of your design.

2. Plot the data as in the basic experiment.

3. Calculate the stiffness-to-weight ratio, k/W.

4. Discuss how the beam could be improved. State the realistic maximum value of k/W that could be expected from your design.

20

TORSION TEST

BACKGROUND

It is found from analysis that in torsion $\sigma_{max} = \tau_{max}$ (applied stress). This has the following implications concerning fracture or yielding occurring first. Assume a material has fracture strength σ_f, and shear yield strength τ_y. Then, there is

(a) elastic deformation only, if $\sigma_{max} < \sigma_f$ and $\tau_{max} < \tau_y$

(b) impending fracture, if $\sigma_{max} \to \sigma_f$ and $\tau_{max} < \tau_y$

(c) impending yielding, if $\tau_{max} \to \tau_y$ and $\sigma_{max} < \sigma_f$

Thus, a tensile failure (b) is possible in a torsion test.

OBJECT

Perform torsion tests of ductile and brittle materials to show the amounts of twisting and the different modes of failure possible.

EQUIPMENT

1. Torsional testing machine with torque and angular twist measuring devices.

2. Safety shield for brittle specimens.

EXPERIMENT

1. Test a few different specimens such as aluminum, low carbon steel, cast iron, plexiglas.

2. Record torque versus angular twist throughout each test, if possible.

REPORT

1. Plot all available torque versus angular twist data on the same diagram.

2. Sketch and discuss the fracture surface of each specimen.

3. Discuss the modes of failure (ductile or brittle).

4. Try to obtain σ_f and τ_y from the torsion tests, and compare these with the data obtained from tension tests of the same materials if possible.

21

STRAIN LINEARITY IN TORSION

BACKGROUND

The shear strain in a cylindrical torsion member varies linearly with the distance from the centerline of the member. Most people easily accept this for cases involving elastic deformations only. Generally, there are serious questions concerning this "theory" for torsion members that have some nonlinear elastic or plastic deformations. An interesting visual proof of this theory can be easily obtained.

OBJECT

Demonstrate the linearity of deformations for small and large angular twists of solid cylindrical rods.

EQUIPMENT

Torsion machine if a metal specimen is used. No machine is needed for weak specimens.

EXPERIMENT

1. Make a smooth cylindrical specimen with a diameter of about 2 cm. Good materials to use are rubber, silicone, mild steel, copper, or aluminum. Drill a very small transverse hole through the center of the specimen.

2. If the specimen is weak enough, twist it with your hands. Look through the transverse hole at various deformations and record your observations.

3. Strong specimens must be twisted in a machine. Again, look through the hole and note any important changes.

4. If a specimen can be spent, test it to fracture, and add to your observations those of the fracture surface and of the transverse hole in its final shape.

REPORT

1. Describe the specimen and the test procedures.

2. State your observations and discuss what they mean for torsion in general.

22

TORSIONAL LOAD CELL, OPTICAL

Apply the concept of the "Beam Load Cell, Optical" (Exp. 12) to make a torque measuring device.

23

TORSION OF THIN-WALLED TUBES

BACKGROUND

The torsion resistance of thin-walled tubes is an interesting problem that can be analyzed rather easily. For a given amount of material in the tube, the shape of the cross section is the most important to consider.

OBJECT

Determine the torsion resistance of rectangular tubes that are of identical material and weight but different in cross-sectional shape. For optional work, also test tubes of other shapes.

EQUIPMENT

1. Gripping and loading fixture to apply torsion loads to the tubes without bending or axial loading.

2. Device for measuring angular twist.

EXPERIMENT

1. Construct thin-walled tubes of paper, wood, plastic, or metal, etc. To make the perimeters the same, $4a = 6b$ (see Figure 11). Use at least these two geometries for the cross sections. Any other shapes must have the same perimeters.

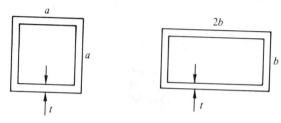

FIGURE 11

2. Apply a small torque to one of the tubes. Measure the angular twist θ.

3. Apply a torque to another tube until its angular twist is also θ. Do this with every tube made of the same material.

REPORT

1. Describe the specimen material, geometries, and the construction technique.

2. Describe the test equipment.

3. Give the results of torque versus angular twist.

4. Calculate the torques in the different tubes, assuming identical shear stresses in them.

5. Compare the theoretical and experimental results of the torsion resistance of different tubes. Discuss the findings and their practical implications.

24

SHEAR CENTER

BACKGROUND

The concept of the shear center is important in dealing with the deformations of certain beams. The possibility of unintentional torsion of a beam can be demonstrated most easily with thin-walled aluminum extrusions.

OBJECT

Experimentally determine the location of the shear center of a small channel-shaped or more complex beam (aluminum extrusions are available in a large variety of shapes). If possible, compare the results with those from calculations.

EQUIPMENT

1. Vise.
2. Weights.
3. Dial gage or long pointer.

EXPERIMENT

1. Devise a grip for the specimen to be loaded as a cantilever beam. A suggested method is to cast it in a short piece of steel pipe using plaster or epoxy as shown in Figure 12.

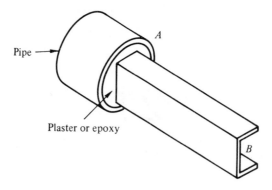

FIGURE 12

2. Grip pipe A in the vise so that the beam is horizontal.

3. Apply a small load at B in such a way that the position of its line of action can be adjusted.

4. Measure the angular twist of end B with respect to A using a dial gage or a long pointer.

5. Find the position of the shear center.

REPORT

1. Describe the beam and the test procedure.

2. Try to justify the experimental result on the basis of theoretical considerations.

3. Discuss the major causes of experimental error.

25

REPEATED SHARP YIELDING

BACKGROUND

The sharp yielding in a mild steel specimen can be observed repeatedly (even hundreds of times) when the so-called strain aging occurs. The return of the sudden yielding depends on plastic straining followed by aging for a sufficient time at a sufficient temperature. This allows diffusion of solute carbon and nitrogen atoms in the steel and results in its strengthening (each yield strength may be a little higher than the preceding one). A simple test with paper clips can demonstrate qualitatively the repeated yielding phenomenon and also allow you to feel the sharp yielding with your fingertips.

OBJECT

Perform torsion of paper clips with repeated strain aging. For optional work, perform a tension test in a testing machine with repeated strain aging.

EQUIPMENT

1. Pliers and candle for the torsion test.

2. Mechanical testing machine with load versus time or load versus deformation recording equipment for the tension test.

3. Propane torch, electric, or radiant heater.

EXPERIMENT

1. Bend a steel paper clip into a U shape. Hold the vertical parts with your fingers and very slowly twist the center part.

2. After each complete relative turn of the ends, heat the center part with the candle for about 30 seconds (longer times are allowed). Let the wire cool before twisting it again. Repeat the process several times or until the wire breaks.

3. Slowly twist the wire (always in the same direction) after each heating and be alert to experiencing the tiny but sudden yielding that occurs in almost every revolution of the permanent twist.

4. In the tension test use any mild steel specimen. Stop the test and heat the specimen with the torch or heater for about one minute or more after each sharp yielding is recorded. Variation of the load on the specimen caused by the heating is allowed. A maximum specimen temperature of about 300°C is sufficient; this is not critical, so temperature measurement is not necessary.

REPORT

1. Describe the test procedures.

2. Describe the observations and enclose any test records.

3. Discuss the practical significance of the observed phenomenon.

26

LIQUID METAL STRAIN GAGE

BACKGROUND

Common strain gages made of metal foil appear very delicate and weak. Yet, there are situations where they are too rigid and strong, or fail too soon from repeated straining. More compliant and more deformable gages are desirable, for example, when testing polymers for automobile tires. A simple device that satisfies these requirements is the unusual liquid metal gage. A thin filament of mercury in a flexible silicone tube is the essence of this gage.

OBJECT

Make a liquid metal gage and calibrate it.

EQUIPMENT

1. Sensitive ohmmeter.

2. Test setup for applying small tensile loads.

3. Micrometer or other device for measuring deformations in rubberlike materials.

1. Construct at least one liquid metal gage. Mold and cure some silicone adhesive or sealer to make a rectangular body with a cylindrical hole through its length as shown in Figure 13. Transparent silicone is best. Inject clean mercury with a hypodermic needle into the hole to make a continuous filament in the center half of the tube (avoid bubbles in the mercury). Insert tight fitting wire terminals to contact the mercury. Glue the terminals in place.

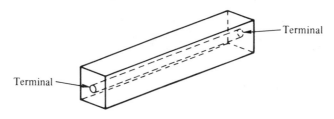

FIGURE 13

2. Test the gage with an ohmmeter for continuity.

3. Stretch the gage slightly (axially) with your hands, and measure the resistance changes.

4. Mount the gage on a thin rubber sheet and grip the latter for tensile loading. The grips should be at a fair distance from the gage.

5. Apply small, known increments of the deformation to the mercury filament, and record the corresponding resistance changes. Unload by increments and again record the resistance changes. Repeat these several times, gradually increasing the maximum deformation from sequence to sequence, even until failure of the gage.

REPORT

1. Describe the materials and procedures used in making the gage. Describe the instruments used in the tests.

2. Plot deformation versus electrical output, and true strain versus electrical output for the gage.

3. Discuss the quality of the gage, and compare it with a common metal foil strain gage if you can.

4. Propose realistic changes to improve your gage.

27

HARDNESS

BACKGROUND

Hardness tests are commonly performed to measure the resistance of materials to permanent indentation, wear, or scratching. Although these tests are cheap and easy to do, the results do not always correlate with other important mechanical properties.

OBJECT

Demonstrate indentation hardness tests on standard testing equipment.

EQUIPMENT

Use as many of the following as possible: Brinell, Rockwell, Shore, and microhardness testers.

EXPERIMENT

Take indentation hardness readings using a few common metals. Observe the deformations.

REPORT

1. Give the data in tabular form.

2. Describe the deformations and discuss the destructive aspects, if any, of each test.

28

BUCKLING

BACKGROUND

The instability of slender columns in compression is easy to demonstrate with small objects, such as yard sticks, axially loaded by hand. In that case, complete failure of the specimen can be prevented by human control in limiting the deformations. Similarly, one must use deformation control and not load control when buckling specimens in testing machines; the idea is to show the onset of instability, not to have catastrophic collapse of the column.

OBJECT

Experimentally determine the buckling loads for uniform cylindrical steel rods of various lengths.

EQUIPMENT

Mechanical testing machine with load measuring device.

EXPERIMENT

1. Make the ends of each specimen pointed to fit in greased sockets on the testing machine. The purpose is to have pinned-end conditions, with a minimum of bending moment applied to the specimen.

2. Gradually increase the compressive load on each specimen and watch for the onset of bending. Record the load at that time.

REPORT

1. Describe the specimens and state the critical loads for them.

2. Plot the critical stress versus the slenderness ratio for the specimens on a single diagram. Show the theoretical values obtained from the Euler equation on the same diagram.

3. Discuss the possible reasons for any differences between the theoretical and experimental results.

29

CREEP OF LEAD

BACKGROUND

Creep in metals is important when the temperature is somewhat above half of the melting point temperature (Kelvin). Thus, lead is excellent for a demonstration since it creeps at room temperature.

OBJECT

For at least one level of load, determine the creep rate and the complete creep curve for lead.

EQUIPMENT

Grips, weights, calipers, timer.

EXPERIMENT

1. Measure the minimum cross-sectional area of the specimen (flat strip) and establish the gage length on a uniform section of it.

2. Apply the load and measure the elongation at reasonable intervals (frequently, if the creep is rapid).

REPORT

1. Plot the complete creep curve, strain versus time, assuming that the initial stress is constant during the test. Show the regions of primary, secondary, and tertiary creep.

2. Determine the creep rate.

3. Determine the stress at fracture. Estimate the complete creep curve for a constant stress that is the average of the initial and final stresses.

30

CREEP OF MAGNETIC TAPE

BACKGROUND

Magnetic tapes are polymers that gradually elongate even under small tensile loads. The elongation changes the apparent frequency of the recorded sound. There are two practical ways to minimize the problem. High quality tapes are prestretched at the factory; this helps because the initial stretching is the most severe under a given load. The other way is to minimize the tension on the tape during transfer from reel to reel. The best machines apply less than 1N tension to the tape.

OBJECT

Determine the creep curves, strain versus time, of a common magnetic tape. Use several loads in the range 0.1 to 10N, and a minimum test time of 10 minutes for each specimen, or 10 minutes at each load when it is increased incrementally.

EQUIPMENT

1. Dead weights.

2. Deformation measuring device that applies little or no force (optical) to the specimens.

3. Micrometer.

EXPERIMENT

1. Design grips for the tape specimens.

2. Measure the cross-sectional area of each specimen before and after the test.

3. Hang the full load on each specimen at once, but try to avoid dynamic loading.

4. Make deflection measurements at reasonable intervals for plotting the creep curves.

5. For optional work, test low and high quality tapes.

REPORT

1. Describe the experimental procedures.

2. State the initial and final stress in each specimen.

3. Give the creep curves in a single diagram.

4. Discuss the results and their relevance to high fidelity recording. State whether the creep causes a decrease or an increase in the frequency of a recorded sound, and give the magnitudes of any such changes on the basis of your tests.

31

ELASTIC AND INELASTIC THERMAL STRAINS

BACKGROUND

Two different strains must be distinguished in a member when its temperature is raised. The strain associated with the volume change depends only on the temperature, and it is generally a recoverable strain. The creep strain depends on the load, temperature, and time, and it is a permanent strain.

OBJECT

Demonstrate the recoverable and permanent strains in a simple elevated-temperature test.

EQUIPMENT

1. Ni-chrome or similar resistance wire specimen.

2. Weight.

3. Variac or other adjustable electric power supply.

4. Ruler or other deflection measuring device.

EXPERIMENT

1. Hang the weight on the straight wire specimen.

2. Send a small current through the specimen. Gradually increase the current until there is a noticeable elongation of the wire. Stop the current at once and let the wire cool. Measure the final deformation, if any. Repeat the procedure a few times, with a slightly higher peak current each time.

3. Repeat the procedure starting with a low current, but allow at least one hour for a creep test. Measure the deformation after cooling. Increase the current for another creep period (same length as for the first creep period). Measure the deformation after each creep period. The creep tests may be continued until the wire breaks.

REPORT

1. Describe the test procedures.

2. State the extent of elastic (recoverable) and inelastic (permanent) strain in each heating period. These should be based on measurements on the alternately hot and cold wire.

3. Discuss the chances of misinterpreting strains in the hot wire and any practical significance of such errors.

32

THERMAL FATIGUE

BACKGROUND

Rapid heating or cooling solids causes thermal stresses in them. Repetitions of these stresses may lead to fatigue cracking.

OBJECT

Apply rapid thermal cycling to several different materials to rank them according to their resistance to thermal fatigue.

EQUIPMENT

1. Electric motor (variable speed is best) with metal wheel and cycle counter.

2. Heater with hot gas flame.

3. Forced cooling air (optional); this may drive the wheel instead of a motor.

EXPERIMENT

1. Make specimens of different materials with identical geometry.

2. Mount the specimens on the wheel and balance the wheel.

3. Aim the flame at the path of the specimens as shown in Figure 14. If there is forced air cooling, aim the air near the flame to cool the specimens immediately after they leave the hot spot.

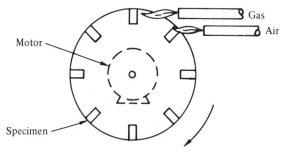

FIGURE 14

4. Stop the test at reasonable intervals (more frequently in the beginning) to look for cracks in the specimens. Record your observations. If there is no cycle counter, calculate the number of cycles of heating and cooling from the time and speed of rotation.

REPORT

1. Describe the specimens and the test setup.

2. Rank the specimens according to their resistance to thermal fatigue.

3. Discuss your observations of the specimens made during and after the test.

33

SIMPLE BENDING FATIGUE TEST

BACKGROUND

Fatigue tests often require complex equipment, but it is possible to perform meaningful tests with one's own hands. The idea is basically the same as in breaking wires by bending them back and forth. Controlling the deformations during bending is the key to calculating the cyclic strain amplitudes and plotting strain versus life curves.

OBJECT

Bend wires against cylindrical mandrels to control the minimum radii of curvature in fully reversed deformations.

EQUIPMENT

1. Micrometer

2. Vise (optional).

3. At least two pairs of steel mandrels for holding the wire specimens and for controlling the deformations. Suggested diameters are shown in Figure 15 (exact values are not important):

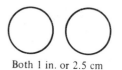

Both 0.5 in. or 1 cm   Both 1 in. or 2.5 cm

FIGURE 15

EXPERIMENT

1. Grip a wire specimen (aluminum, copper, or iron) between the mandrels, as shown in Figure 16. Ideally, the wire should be held firmly, without indenting it by the mandrels. This can be tricky, so one workable solution is described here. The thin insert I under turns 1 and 2 should not be too brittle or too pliable (many metals, paper, tape, etc. are OK). S is the free end of the wire specimen. Bolts B (or a vise) are tightened, compressing turns 1 and 2 of the wire. The tightening should stop when lines L_1 and L_2 of the two mandrels just touch S, without indenting it.

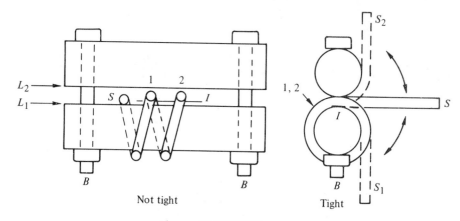

Not tight   Tight

FIGURE 16

2. Bend S back and forth between the approximate extreme positions S_1 and S_2 until the wire breaks or time and patience run out.

3. Test at least three specimens of a given wire for each mandrel size.

REPORT

1. Describe the experimental setup.

2. Calculate the strain amplitude in the wire (from the original or mean position of S to the extreme S_1 position). This should be based on the dimensions of the wire and the mandrel. Assume that the strain varies linearly from zero at the neutral axis (centroidal) of the wire.

3. Plot on a single diagram the strain amplitude versus number of cycles to failure for each specimen.

4. Discuss the results, the causes of scatter, and the probable stress levels in the materials tested.

74

34

STRESS-STRAIN CURVE CHANGES

BACKGROUND

For most metals, the stress-strain curve is not unique. It can be altered by thermal treatments or plastic deformations. A few of the possible changes can be easily demonstrated with very little equipment.

OBJECT

Observe the large changes in stress-strain response in copper or aluminum wire caused by alternate annealing and cold-working.

EQUIPMENT

1. Thick copper or aluminum wire (several mm dia.).

2. Fireplace or furnace.

3. Vise or mandrels as in "Simple Bending Fatigue Test" (Exp. 33).

EXPERIMENT

1. Slowly bend the wire back and forth in fully reversed deformations as in the "Simple Bending Fatigue Test." Apply ten complete cycles. Record what your hands feel concerning the deformation resistance of the wire from cycle to cycle.

2. Heat the wire to red hot for a few minutes.

3. After the wire has cooled, repeat the ten cycles of bending and the recording of the behavior.

4. Continue the alternate heating and deformation treatments five to ten times.

REPORT

1. Sketch the estimated stress-strain curve for the original wire during the first bend. If the wire was extremely soft at first, the stress-strain curve has no elastic (straight line) portion; assume that the yield strength at 0.2% offset is about 3 ksi = 20 MPa.

2. Sketch on the same diagram the estimated curve for the tenth cycle.

3. Sketch on the same diagram the estimated curves for the first and tenth cycles after each heating treatment.

4. Compare the curves and discuss your observations.

35

CYCLIC PLASTIC HEATING

BACKGROUND

Most people have experienced the heat that is generated when wires are bent rapidly several times. The heating is from the plastic deformation of the metal. In certain mechanical loading systems, the cyclic plastic heating can melt large metal parts. Even in less severe cases, the temperature changes are indicative (using special instrumentation) of the fatigue damage process which is closely related to the cyclic plastic straining.

OBJECT

Demonstrate the cyclic plastic heating in a mild steel bar loaded as a cantilever beam.
Note: This demonstration can be very impressive. A small group of students has won an award with it at an Engineering Exposition at the University of Wisconsin.

EQUIPMENT

1. Mechanical testing machine for continuous cyclic loading at the rate of 10 Hz or more.

2. Thermocouple and temperature indicating instrument; stroboscope light (optional).

EXPERIMENT

1. Grip a flat bar specimen S to the actuator A of the testing machine as shown in Figure 17. Clamp a small weight W to the free end of the specimen.

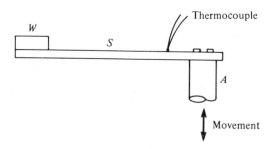

FIGURE 17

2. Try specimens of various lengths, or vary the weight to obtain fracture of the specimen in 5 to 10 minutes.

3. Record temperatures if possible.

4. Observe the deflection of the beam with the aid of the stroboscope if available.

5. Drip water or liquid nitrogen on the specimen during cyclic loading. This is especially effective in conjunction with the stroboscope lighting.

REPORT

1. Describe the specimens and the test setup.

2. Try to correlate specimen temperatures with number of cycles to failure.

3. Try to estimate the strains in the specimens from their deflection curves.

4. Describe and discuss any unexpected behaviors.

36

RANDOM LOADING

BACKGROUND

Even in a given point in a member, stress and strain are seemingly unrelated when the deformations involve plasticity. This has particular significance in the analyses of fatigue damage accumulation, service load histories, and the laboratory simulation of service load spectra.

OBJECT

Observe the differences between stress and strain in a specimen under random loading.

EQUIPMENT

1. Closed-loop, servo-controlled testing machine with stress and strain measuring devices.

2. A two-pen strip chart recorder, or two with single pens.

3. X-Y plotter, or storage oscilloscope for X-Y recording (optional).

4. Random function generator (optional).

EXPERIMENT

1. Determine the maximum desirable stress and strain for a mild steel or aluminum specimen, and try not to exceed these levels.

2. Vary at random the amplitude and mean level of the chosen controlled quantity. Manual control (by one or two people) is acceptable. Record both the controlled and dependent quantities on the strip chart and X-Y recorders.

3. Use strain control and then stress control on the same specimen.

REPORT

1. Describe the specimen's known relevant mechanical properties.

2. Discuss the most noteworthy features of the recordings.

3. Evaluate the differences between the effects of the two control conditions.

37

ACOUSTIC CRACK DETECTION

BACKGROUND

In the old days, cracks in railway wheels were detected by hitting them with a small hammer and listening to the sound. The ringing sound depended on the size of cracks and could be distinguished from the sound of "good" wheels.

OBJECT

Perform simple experiments to determine the smallest notch depth that can be detected with the human ear.

EQUIPMENT

1. A piece of steel or aluminum.

2. Tape recorder.

EXPERIMENT

1. Suspend or clamp the test piece. Hit it with a small metal object and record the sound.

2. Saw a very small notch (a fraction of a mm deep) in the test piece. Hit it exactly as you did before the notching and record the sound.

3. Progressively deepen the notch by very small amounts and record the sound for each notch depth.

4. Have someone play back the various sounds at random. Relying on the sounds alone, try to pick out the unnotched piece and the one with the smallest notch.

REPORT

1. Describe the details of the equipment used and the experimental procedures.

2. State the results and the expected range of error.

3. Discuss how the sensitivity (detection of smaller notches) of the experiment could be increased and the chances for error reduced.

4. Discuss how the results might change if the test piece were smaller or larger than the one used.

38

FRACTURE OF ALUMINUM FOIL

BACKGROUND

Fracture load or stress of a specimen is not a material property. It depends on the geometry of the specimen, especially on the length of cracks or defects. You can demonstrate this, and the use of the stress intensity factor, in a simple test.

OBJECT

Perform fracture tests on notched, thin specimens of aluminum foil.

EQUIPMENT

1. Weights or mechanical testing machine with load measuring device.

2. Micrometer and ruler.

EXPERIMENT

1. Cut a few specimens from aluminum foil, each about 8 cm X 40 cm in size.

2. Make grips to transfer the tensile load fairly uniformly at each end of a specimen (over the 8 cm width). One good method of gripping is to wrap and tape the foil on

cylindrical rolls, and apply the load through the latter. Loads between 20 and 100N are reasonable.

3. Cut a notch into the center of the strip with a razor blade or a sharp knife. The notch should be perpendicular to the loading axis.

4. Gradually increase the notch length symmetrically in both directions until the strip fractures.

5. Use the notch length at the onset of fracture to calculate K_I, the opening mode stress intensity factor, from

$$K_I = \sigma \sqrt{w \tan \frac{\pi c}{w}}$$

where σ = average stress

w = width of the strip

$2c$ = notch length

6. Use another strip with a given length of notch, and increase the load from zero until fracture of the strip. Recalculate K_I from the results of this test, and compare with the previous ones.

7. Examine the cut and fractured surfaces with a magnifying lens of 10 to 50X.

REPORT

1. Describe the specimens and the test setup.

2. State the K_I values and discuss the possible reasons for any significant differences.

3. Describe the observed features on and near the fracture surfaces.

39

NOTCH STRENGTHENING

BACKGROUND

Ductile metals may fracture in a brittle mode if the maximum shear stress is less than the yield strength in shear. Such a condition can exist in notched members where even uniaxial loading causes triaxial states of stress. The result is that the apparent yield strength increases (and this appears good), but a ductile metal acts as if it were brittle (and this *is* bad).

OBJECT

Compare the yielding and fracture behaviors of notched and unnotched specimens of the same mild steel.

EQUIPMENT

Mechanical testing machine with load and deformation measuring and recording devices.

EXPERIMENT

1. Make a circumferentially grooved (sharp groove) and an unnotched tension specimen from the same low carbon steel. Measure the minimum diameters in the specimens.

2. Perform the tension tests. Record load versus time or load versus deformation.

REPORT

1. Describe the specimens and the test setup.

2. Discuss the data, especially the yield, ultimate, and fracture strengths.

3. Describe the main features of the fracture surfaces, and determine the percent reductions of the areas.

40

MULTIPLE NOTCHES

BACKGROUND

Notches cause stress concentrations and these tend to lower the fracture strength of a member as a whole. Two or more notches near each other often worsen the situation, but surprisingly, they may be beneficial in some cases. The reason for the latter is that more notches may enhance the energy dissipation in plastic zones at the notch roots.

OBJECTS

Attempt to show beneficial and harmful effects caused by various geometric arrangements of identical holes in notched aluminum foil specimens.

EQUIPMENT

1. Dead weights or a mechanical testing machine to apply small tensile loads.

2. Optical microscope (optional).

EXPERIMENT

1. Cut a few dozen identical specimens from thin aluminum foil. Make a single, sharp notch of identical dimensions in each as shown in Figure 18. L should be large enough for gripping the specimens.

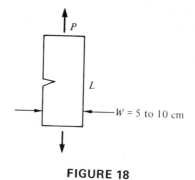

FIGURE 18

2. Test several of the edge-notched specimens to determine the average fracture strength.

3. Use a pin less than 1 mm in diameter to make two identical holes in each of the remaining specimens. Follow the two basic patterns shown in Figure 19, but you are free to choose the locations and spacings of the holes. Aim to arrange the holes to maximize or minimize the fracture strength.

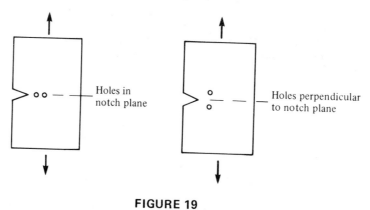

FIGURE 19

4. Look at the notch regions for evidence of plastic deformations.

REPORT

1. Describe the details of the equipment used and the experimental procedures.

2. State the results and the expected scatter.

3. Theorize about the results.

4. Describe the notch regions of the broken specimens.

5. Theorize about the possible effects of changing the specimen geometry in one or more ways.

41

STRAIN CONCENTRATION
IN A RUBBER SHEET

BACKGROUND

The strains in a notched, soft rubber sheet can be so large that the effect of the notch is visible. The localized deformations can be demonstrated even with crude measurements.

OBJECT

Determine the approximate strain distribution in a uniformly loaded, notched rubber sheet.

EQUIPMENT

1. Mechanical testing machine or weights.
2. Rubber sheet specimen with grid lines (see Figure 20).
3. Calipers.

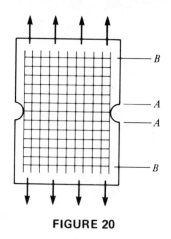

FIGURE 20

EXPERIMENT

1. Check the distances of lines A-A and B-B at a few places before loading the specimen.

2. Apply a tensile load and measure the new spacing of lines A-A and B-B across the specimen.

3. Check a few spacings after unloading the specimen.

REPORT

1. Plot the strain distributions across the minimum section, using both the short (A-A) and the long (B-B) gage lengths.

2. Discuss the possible reasons for any sizable difference between the two strain plots.

3. Estimate the true strain distribution across the minimum section, and give your reason for it.

42

STRESS CONCENTRATIONS
IN A METAL

BACKGROUND

Discontinuities in members cause nonuniform distributions of stresses and strains. These are important in the analysis of critical regions and loading, and in efforts to prevent failures. Experimental investigations of stress concentrations are based on localized strain measurements.

OBJECT

Determine the approximate stress distributions and stress concentration factors in an axially loaded bar with several different notches.

EQUIPMENT

1. Mechanical testing machine with load measuring device.

2. Aluminum or steel notched bar with strain gages (see Figure 21 for recommended geometry and gage locations).

3. Strain readout or recording device.

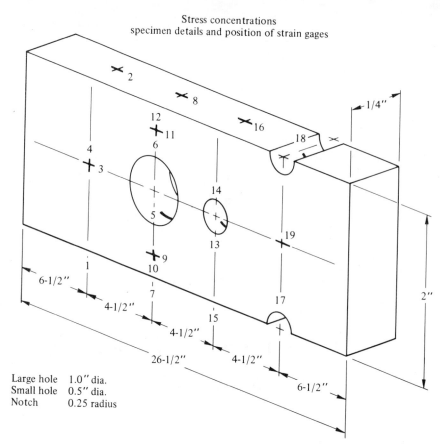

Stress concentrations
specimen details and position of strain gages

Large hole   1.0″ dia.
Small hole   0.5″ dia.
Notch   0.25 radius

FIGURE 21

EXPERIMENT

1. Load the bar in tension to produce only elastic deformations in it.

2. Record the strains at all gage locations.

REPORT

1. Describe the specimen.

2. Calculate the elastic modulus, Poisson's ratio, and the stresses at all gage locations. Where the stress is uniaxial, use $\sigma = E\epsilon$. At gages 9, 10, 11, and 12 there are biaxial stresses (longitudinal and transverse) because of the mutual constraint of neighboring notched and unnotched segments of the bar. Here the generalized Hooke's law must be used. For example, σ_9 and σ_{10} can be obtained from

$$\epsilon_9 = \frac{1}{E}(\sigma_9 - \mu\sigma_{10})$$

$$\epsilon_{10} = \frac{1}{E}(\sigma_{10} - \mu\sigma_9)$$

3. Determine the stress concentration factors for the notches.

4. Plot the stress distribution for each reduced cross section. Integrate the stresses over each cross section and compare the resulting total force with the applied load. Discuss the possible reasons for any discrepancy.

STRESS CONCENTRATIONS DATA

Strain indicator equipment _____

Initial strain reading _____

Load on specimen _____

Gage No.	Strain reading	Strain increment	Actual strain	Stress
1				
2				
3				
4				
5				
6				
7				
8				
9				
10				
11				
12				
13				
14				
15				
16				
17				
18				
19				
20				

43

PLASTIC ZONES
IN A NOTCHED PLATE

BACKGROUND

Localized plastic deformations in critical regions of members (flaws, machined holes, etc.) are important in investigating the failure mechanism and in efforts to prevent failure. In some cases, the plastic zone is clearly seen with the naked eye. For example, in hot-rolled mild steel the brittle oxide layer flakes off in the regions of plastic deformation.

OBJECT

Make detailed visual observations of the size and shape of plastic zones in mild steel.

EQUIPMENT

Mechanical testing machine with grips for flat plate specimens.

EXPERIMENT

1. Make at least one notched specimen from a thin sheet of hot-rolled mild steel. Various notch sizes and shapes are acceptable, including center hole, single or double edge notches.

2. Gradually increase the tensile load on the specimen until the first signs of yielding appear. Record your observations of the plastic zone(s).

3. Increase the load in steps, and each time record your observations.

REPORT

1. Describe the specimen material and geometry.

2. State your observations with the aid of sketches.

3. For optional work, analyze the early stages of yielding using stress concentration factors. Evaluate the calculations and the observations.

44

ICE CUBE CRACKLING

BACKGROUND

You have probably heard the crackling of ice cubes while making cold drinks. The phenomenon involved is interesting, yet most people could not tell how or why the crackling occurs. The practical significance of this extends beyond kitchens and bars. The improvement of methods of breaking ice in harbors and waterways depends on the understanding of the mechanical behavior of ice.

OBJECT

Attempt to cause crackling of ice cubes by rapid temperature change.

EQUIPMENT

1. Freezer.

2. Heater.

3. Thermometer or thermocouple with meter or recorder.

EXPERIMENT

1. Make essentially flaw-free, unnotched ice cubes.

2. Drop ice cubes in water at various temperatures, lift them out quickly and observe the crackling of the cubes.

3. For optional work, test ice cubes that have sharp but shallow notches scratched on the surface.

REPORT

1. Describe the specimens and the experimental procedures.

2. State the observations and the test results.

3. Theorize about the behavior of the ice cubes.

4. Speculate about any practical significance of your test results for the breaking of thick ice on a large body of water.

45

LIQUID METAL EMBRITTLEMENT

BACKGOUND

Some solid metals can be severely embrittled when they are in contact with certain liquid metals. Such a possibility must be considered in transporting or using liquid metals, for example in heat exchangers using sodium.

OBJECT

Demonstrate that mercury at room temperature can embrittle brass.

EQUIPMENT

Vise, or pliers.

EXPERIMENT

1. Use brass wires with a diameter of about 2 to 3 mm for specimens. Take a few of these and bend them back and forth until fracture. Count the cycles to failure. Try to have the same minimum radius of curvature in each bend.

2. Take identical wires and carefully clean the intended test sections. These parts should be free of thick oxide layers, oil, grease, paint, water, or dirt. Immediately after cleaning, dip these sections in mercury for a few minutes. The mercury should wet the clean brass.

3. Perform the bending tests on the latter specimens, and count the cycles to failure.

REPORT

1. Describe the test procedures.

2. Discuss the results and their implications, including any generalizations that may be reasonable to other solid-liquid couples.

46

COMBINED STRESSES

BACKGROUND

Frequently, a given member is loaded in bending and torsion simultaneously. The analytical or experimental stress analysis of such a member is straightforward if the deformations are entirely elastic.

OBJECT

Demonstrate the superposition of stresses in a cantilever beam loaded in bending and torsion.

EQUIPMENT

1. A tubular aluminum or steel beam with a three-element strain gage rosette.

2. Strain gage readout equipment.

3. Weights.

EXPERIMENT

1. Clamp the beam horizontally as shown in Figure 22 with the rosette at A. One gage should be mounted longitudinally, another transversely with respect to the centerline of the beam.

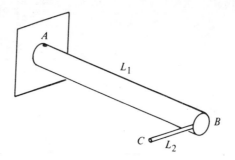

FIGURE 22

2. Apply the load at B (bending only), and record the strains.

3. Apply the same load at C (bending and torsion), and record the strains.

REPORT

1. Describe the details of the beam and the loading.

2. Compute the normal stress for pure bending ($\sigma = \frac{Mc}{I}$), and plot stress versus longitudinal strain and stress versus transverse strain. From this plot determine the modulus of elasticity. Also calculate Poisson's ratio.

3. For bending and torsion (load at C), determine the experimental and theoretical principal stresses σ_1 and σ_2 and their directions at point A.

4. Compare the experimental and theoretical results, and discuss the possible reasons for any significant differences.

47

IMPACT TOUGHNESS

BACKGROUND

Impact toughness values obtained at different specimen temperatures are useful for comparing materials in the design process. The Charpy test is the most common for such a purpose. It is a relatively cheap and simple test for the determination of the ductile-brittle transition temperature. The impact energy values cannot be used directly in design.

OBJECT

Perform impact tests at three or four different temperatures on low, medium, and high carbon steel specimens.

EQUIPMENT

1. Impact tester of the pendulum type.

2. Heater and freezer.

EXPERIMENT

1. Learn to calibrate the machine to compensate for friction.

2. Test the specimens at the following temperatures if possible: room temperature; near the boiling point of water; freezer temperature; dry ice or liquid nitrogen.

REPORT

1. Plot all the data on the same diagram of impact energy versus temperature. Connect the points with a reasonable, smooth curve for each metal.

2. Discuss the results and the possible reasons for any scatter or unreasonable positions of data points.

48

PRIMITIVE CHARPY TEST

BACKGROUND

The standard Charpy test is quite simple and requires relatively inexpensive equipment. However, it can be envisioned that test results are needed quickly and there is no access to a testing machine. Test specimens are fabricated in a few hours, but what is the best home-made test setup?

OBJECT

Test standard Charpy specimens in the simplest way you can devise.

EQUIPMENT

Anything except standard Charpy testers. The latter are allowed only for comparison of the test results if desired.

EXPERIMENT

When the test setup is completed, break a few specimens to check the operation and repeatability of the procedures.

REPORT

1. Describe the principles of your design.

2. State the test results and indicate whether they may be acceptable or not.

3. Discuss how your tests could be improved.

49

DEFORMATIONS FROM
IMPACT LOADING

BACKGROUND

The deformation of a structure is larger when a weight is dropped on it than with a static weight of the same size. For an elastic structure the problem can be reduced to the deformation of an equivalent spring. The static and dynamic deflections are related in a simple mathematical expression.

OBJECT

Measure and calculate the deflections of a cantilever beam under static and dynamic loading.

EQUIPMENT

1. Fairly flexible steel or aluminum tube, about 3 m in length and 2 cm in diameter.

2. Weight and holder.

EXPERIMENT

1. Clamp the tube horizontally, with the weight holder at its free end. The tube may deflect considerably under its own weight.

2. Attach a string to the tube as shown in Figure 23. The motion of the tube should not be hindered significantly.

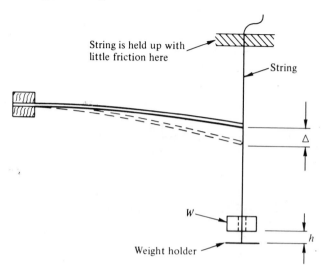

String is held up with little friction here

String

Δ

W

Weight holder

h

FIGURE 23

3. Slowly place W on the holder and measure Δ by how much string was pulled down.

4. Suddenly release W on the holder with $h \approx 0$. Measure Δ.

5. Drop W on the holder from increasing heights h. Measure Δ in each case.

REPORT

1. Calculate the theoretical deflections for each loading.

2. Plot the measured and calculated values of Δ versus h on the same diagram.

3. Discuss the reasons for any discrepancy in the results.

50

TESTING MACHINE RESPONSES

BACKGROUND

It would be ideal to have mechanical tests of materials using a wide range of stress or strain amplitudes, wave forms, and loading rates. Unfortunately, the inertias, deformations, and other characteristics of testing systems (and of specimens) often place severe limitations on what can be achieved in practice. Even the best machines, those using closed-loop servo-control, have these limitations.

OBJECT

Demonstrate the limitations in the meaningful measurement of stress-strain response of a specimen tested in a modern machine.

EQUIPMENT

1. Closed-loop, servo-controlled testing machine with a sine, square, and triangle function generator.

2. Stress and strain measuring and recording apparatus.

EXPERIMENT

1. Use a smooth specimen of mild steel or aluminum. Limit the loads and deformations so that the specimen is not likely to fail through the various parts of the test.

2. Set a constant frequency of cyclic loading between 1 and 5 Hz. Apply a low amplitude sine wave in stress or strain control. Gradually increase the amplitude to the maximum and always record stress and strain. Repeat the process using the other wave forms.

3. The same as in (2), but with the amplitude fixed at about one half of the maximum, and the frequency varied from low to high.

REPORT

1. Describe the specimen and the important nominal characteristics of the testing machine.

2. Analyze the results and state the most important limitations observed.

3. Discuss how the limitations would change for a more or less compliant specimen than the one tested.